ERIE SHOVEL

PHOTO ARCHIVE

Iconografix continuously seeks collections of archival photographs for reproduction in future books. We require a minimum of 120 photographs per subject. We prefer subjects narrow in focus, i.e., a specific model, railroad, racing venue, etc. Photographs must be of high-quality, suited to reproduction in an 8x10-inch format. We willingly pay for the use of photographs.

If you own or know of such a collection, please contact: The Publisher, Iconografix, PO Box 609, Osceola, Wisconsin 54020.

ERIE SHOVEL
PHOTO ARCHIVE

Photographs from the Erie Steam Shovel Company Archives of the Historical Construction Equipment Association

Edited with introduction by
Donald W. Frantz

Iconografix
Photo Archive Series

Iconografix
PO Box 609
Osceola, Wisconsin 54020 USA

Books in the Iconografix *Photo Archive Series* are offered at a discount when sold in quantity for promotional use. Businesses or organizations seeking details should write to the Marketing Department, Iconografix, at the above address.

Library of Congress Card Number 96-78346

ISBN 1-882256-69-7

97 98 99 00 01 02 03 5 4 3 2 1

Printed in the United States of America

PREFACE

The histories of machines and mechanical gadgets are contained in the books, journals, correspondence, and personal papers stored in libraries and archives throughout the world. Written in tens of languages, covering thousands of subjects, the stories are recorded in millions of words.

Words are powerful. Yet, the impact of a single image, a photograph or an illustration, often relates more than dozens of pages of text. Fortunately, many of the libraries and archives that house the words also preserve the images.

In the *Photo Archive Series*, Iconografix reproduces photographs and illustrations selected from public and private collections. The images are chosen to tell a story—to capture the character of their subject. Reproduced as found, they are accompanied by the captions made available by the archive.

The Iconografix *Photo Archive Series* is dedicated to young and old alike, the enthusiast, the collector and anyone who, like us, is fascinated by "things" mechanical.

ACKNOWLEDGMENTS

I would like to thank the Bucyrus-Erie Company for providing historical information contained in this book's introduction and acknowledge, as a reference work, the book *Designed for Digging: The First 75 Years of the Bucyrus-Erie Company,* Williamson and Myers, Northwestern University Press, 1955.

I would like to thank J. Keith Haddock and Ron Ketron for the use of several photos from their private collections. The bulk of the photos used in this book are courtesy of the Historical Construction Equipment Association (HCEA) Archive.

The HCEA is a registered, nonprofit organization founded in 1986 "dedicated to preserving the history of construction, surface mining, and dredging equipment, including various manufacturer's archives, histories of product development, descriptive literature, technical data, photographs, memorabilia, and, of course, the machines themselves."

The HCEA is the only organization in the world dedicated to preserving the history of construction equipment. The HCEA publishes a quarterly magazine, *Equipment Echoes,* covering events, historical equipment news, and educational articles. The HCEA sponsors an annual convention, held in various locations in the US, that attracts hundreds of equipment exhibits and spectators by the thousands. Here, historical equipment and construction techniques are actively demonstrated.

The HCEA plans to build a world-class museum, a place to preserve historically significant construction equipment and to restore, display, and interpret these machines for the education and enjoyment of all visitors for generations to come.

The HCEA is honored to be the only organization designated by the Construction Industry Manufacturers Association (CIMA) as an "official archivist of the construction industry." The HCEA Archive provides a service to both the manufacturers of construction equipment and the general public by preserving, cataloging, and making available for research in one location, all the archival material associated with the industry. This service is duplicated nowhere else in the world.

For more information on the HCEA, please contact:

HCEA
PO Box 328
Grand Rapids, Ohio 43522 USA

J. J. Thomson Company's ERIE Gas+Air shovel on a cellar job. The unique Gas+Air used compressed air-powered engines very similar to the steam engines used on the ERIE steamers. The crowding and swinging functions were performed by air pressure, while hoisting, booming, and traveling were powered mechanically.

INTRODUCTION

The history of the ERIE steam shovel can be traced back to a chance meeting between two men who were in the right place at the right time. A.C. Vicary, salesman "extraordinaire" for the Thew Automatic Shovel Company of Lorain, Ohio, struck up a conversation with Fred McGrier, Vice-President of the Ball Engine Company of Erie, Pennsylvania, while both were riding on a train. After eight years of selling shovels throughout North America, Vicary had very definite ideas about both steam shovel design and the market potential of the machines. Recognizing the diminishing need for steam engines as a source for industrial power, McBrier's family-owned company was looking for a new line to manufacture. Vicary's offer to "design a shovel and to sell all the shovels the company could build" was eventually accepted by the Ball Engine Company, thus thrusting them into the small excavator market. After hiring Vicary, and with help from several others who had special talents, early in 1915 the Erie Type B ¾-cubic yard shovel was introduced. Another model, the ½-cubic yard Type A, was added to the line in 1916.

From the beginning, it was proposed to make the Ball Engine Company "the Ford Motor Company of the excavator industry". By concentrating on only two models and stressing the interchangeability of repair parts, machines could be manufactured quickly from parts that had been batched in quantity, and machines could be serviced in the field more quickly from this same stock of parts. By building a simple, easy-to-operate, roomy machine, with standards of high quality, and through an aggressive sales force, the ERIE shovels gained a large share of the ¾-cubic yard shovel market, and soon became the industry standard.

After selling its stationary steam engine business in 1920, the Ball Engine Company concentrated its efforts on what had become its principal product, steam shovels. In 1922, the company's corporate name was changed to the Erie Steam Shovel Company.

After years of resistance to changing their shovel, in 1923 the Erie Steam Shovel Company realized that their very successful Type B steam shovel had about run its course. The realities of the then current shovel market were finally recognized by the management of the company. These were: the increasing popularity of gasoline-powered excavators; contractors wanted larger shovels than the Type B's ¾ cubic yard capacity; Erie's crawler mounting and steering design were out of date; and the Type B was not easily adaptable to dragline work.

Erie tackled these challenges with the development of two "new" excavator models. The first, introduced in late 1925, was the ERIE B-2 Dreadnaught steam shovel. The Type B-2 was nothing more than a one-cubic yard version of the Type B that it replaced, and was only intended as a stop-gap solution to the increasing popularity of gasoline-powered excavators.

The second new machine, also introduced in 1925, was the novel ERIE GA-2, a one-cubic yard "gas-plus-air" excavator. This novel machine combined a gasoline engine, an air compressor, and air-operated engines—an arrangement that was unique in the excavator industry. Erie felt strongly that this combination would give the shovel owner the best of both worlds. Gasoline power eliminated the fuel and water problems inherent with steam shovels, while the compressed air-operated engines retained all of the desirable operating characteristics of the steamers. Unfortunately, the Type GA-2 excavator did not achieve anything near the success achieved by the Type B steam shovel. The GA-2's high initial cost, and the greater amount of skill and knowledge needed to keep the machine operating properly, were major limitations to its popularity. For the first few years of its production, the GA-2 was outsold by the steam shovels it was designed to replace!

In 1926, the Erie Steam Shovel Company was reorganized, eventually coming under control of the Erie Investment Company. Two of the principal directors of the new board were also directors of the Bucyrus Company of Milwaukee, Wisconsin, then, arguably, the largest manufacturer of mid- to large-size excavators in the world. Their involvement with Erie eventually paved the way for consolidation of the two companies. The Erie machines filled Bucyrus Company's need for small excavators, which they did not manufacture. Erie's worldwide sales network was also a welcomed addition.

Following the formation of Bucyrus-Erie Company in 1928, the combined product line still included the Types A, B2, and GA-2. By the time of the merger, Erie had finally introduced an entirely new chain-driven crawler mounting, a basic design that B-E used for about 45 years! After phasing out the steamers, the Type GA-2 was eventually joined by the Type DA-2, an Atlas diesel-powered excavator, and the Type GA-3, a larger, 1¼-cubic yard excavator. Although they were advertised until 1935, The Gas+Air machines were eventually replaced by a varied list of gasoline and diesel-powered Bucyrus-Eire machines, thus ending the lineage of excavators begun twenty years earlier by the Ball Engine Company.

ERIE Type A steam shovel belonging to Chas. Fry Construction Company, Erie, Pennsylvania. Frost in this cellar ran 10 to 20 inches thick, but reportedly the ERIE broke it up fairly well. The man under the dipper was A. C. Vicary, the "father" of the ERIE Shovel.

Another view of the Chas. Fry Construction Company machine. On some days, even before the contractor started using dynamite to break up the ground, this Type A loaded out as much as 300 cubic yards.

Type A shovel owned by Johnson & Kearns, Dubois, Pennsylvania. The operator was making a difficult cut for a 20-foot wide street. He was working downhill, on a 15 percent grade, and was loading wagons in front of the shovel. Note the cleats bolted to the traction wheels.

Type A shovel, owned by Antonio Colarusso, Hudson, New York, quarried rock for the Knickerbocker Portland Cement Company.

Sowerbutt Quarries was the proud owner of this Type A. They bragged that the ERIE, while working in a seven-foot bank of overburden, “did more stripping in five days than we accomplished last fall in six weeks with eight men and four carts!” The small, ½-cubic yard Type A often replaced hand labor.

Type A shovel owned by Schreier of Buffalo, New York. This deep excavation required loading wagons on an 11-foot high bank.

Type A owned by C. C. Kilby of Hartford, Connecticut. On this fine level grading, the operator worked without any pitmen. The operator did all the finishing with the dipper!

Sweeney Bros. Contracting of Scranton, Pennsylvania, involved in what was an increasingly common application for steam shovels in the early part of the century—trenching. Since the shovel had to constantly move toward its work, it was necessary for the machine to travel over the completed ditch, something that sometimes ended in a disastrous cave-in.

Another view of the Sweeney Bros. machine. The Type A shovel was equipped with a special 19-foot long dipper made for sewer and ditching work. The bucket was narrower than standard and had teeth at the corners to clean sidewalls. The added dipper length drastically cut down on bucket clearance.

Type A Erie shovel owned by Hamilton County, Chattanooga, Tennessee. The attendant water wagon is a reminder of the steam shovel's constant thirst for water.

Another view of the Hamilton County machine. The bottom dump wagons were double hitched, but there is nothing to indicate whether they were horse or crawler tractor-drawn.

A new Type A shovel strutted her stuff behind the Ball Engine Company factory in Erie, Pennsylvania. This may have been the first Type A to be equipped with ERIE's "caterpillar-type" mounting.

The Type A was the second machine introduced by the Ball Engine Company. First offered in 1918, it was preceded by ERIE's flagship machine, the Type B, which was introduced in 1915.

The operator of the new Type A tested its gradeability. One of ERIE's manufacturing advantages was the use of some common parts between the Type A and the Type B machines.

Type A shovel owned by E. N. Wiegel, Little Rock, Arkansas. After moving between 300 and 400 tons of rock per day for six months, the dipper on this machine was practically worn out.

This Type A shovel, owned by Dixie Brick & Tile Company, Columbus, Georgia, excavated clay for the brick plant. This is a good view of the operator at his controls. At that time, a seat for the operator was an option. The brake pedal for the hoist required that the operator spend much of his day standing on one foot!

Type A shovel owned by C. E. Redington of Greensburg, Indiana. On the day this photo was taken, the ERIE handled 352½ cubic yards of blasted limestone in 9 hours and 27 minutes.

Type A shovel, owned by the Troy Public Works Company, Utica, New York, excavated for a new power plant for Henry Ford & Son, Inc.

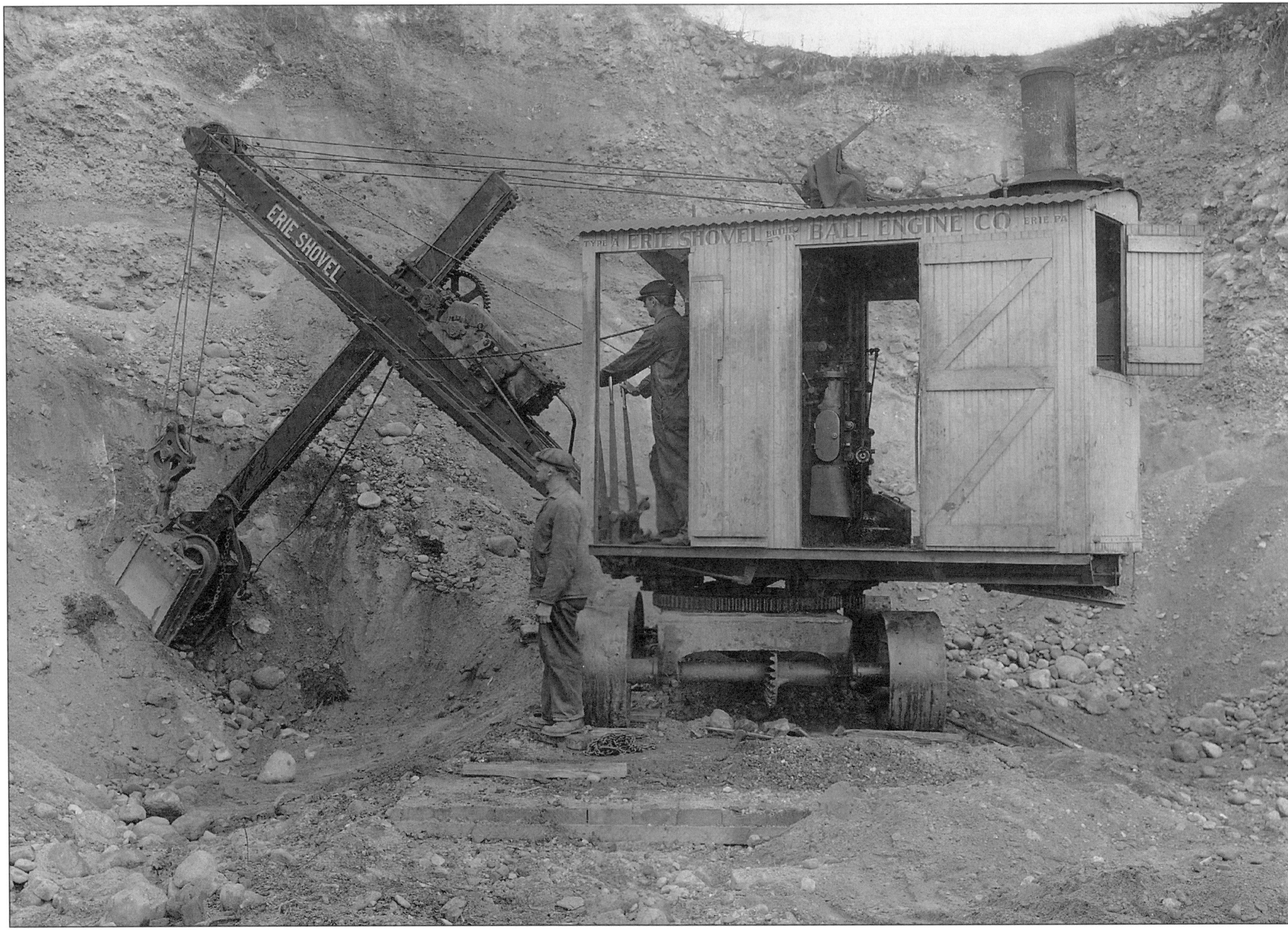

P. Cogger of Lowell, Massachusetts, producer of crushed stone, gravel and sand, boasted that this Type A easily handled 300 cubic yards of gravel a day—50 percent more than he had expected!

Sidehill cut for a Type A, owned by Chas. Spindler, Kingwood, West Virginia. This railroad siding excavation was 18 feet deep, 11 feet of which was rock. The contractor claimed to have moved 350 cubic yards in a day.

New Type A crane being tested before shipment. The company stated, "This speedy little crane carries a 26-foot boom, which is set at the most advantageous angle by means of the boom hoist. The A ERIE Crane operates a ½-cubic yard clam-shell with excellent results."

Truck frame of a Type A traction shovel. ERIE claimed that the combination of great strength and least possible weight was accomplished by means of a three-quarter box section, which carried the strains around the outside of the frame.

Another view of the Type A truck frame. Called "indestructible" by ERIE, the company claimed that because it was a one-piece casting, they never had the slightest problem with the truck frame on any Type A or Type B machine.

New Type A shovel. The dipper and boom were made of selected white oak and clad with steel. Metallurgists could not yet produce a steel capable of withstanding the shocks and boom twisting that were incurred in shovel work.

Winter of 1922 found this Type A shovel, owned by C. C. Kilby, shoveling snow in Hartford, Connecticut. Notice the row of young, would-be shovel runners that watched in awe. Crawler mountings for the Type A seem to have been rare.

Type A shovel on caterpillar-type mounting, owned by Raterman Building & Contracting Company, St. Louis, Missouri, excavated the basement for St. Mary's Hospital. It was always desirable to keep the wagons out of the excavation, and the ERIE's high lift helped get the job done.

ERIE Type A "Tunnel Mucker". Equipped with a special 14-foot 6-inch boom and 10-foot 2-inch dipper arrangement, the small ERIE was readily adaptable to the confines of underground excavation. The chains were hooked to cutting dogs. While traveling, the operator could swing the upper works a little to the left or right, to disengage the travel to the appropriate track and, thus, control the steering.

The Type A "Tunnel Mucker" was powered by compressed air, rather than steam. The exclusion of the boiler reduced the rear swing radius to just 5-feet 6-inches. When operated steadily, the "Tunnel Mucker" consumed about 500 cubic feet of air per minute. Air was fed up through the center pin. Note that the word "SHOVEL" had been dropped from the boom plate and only the name "ERIE" was used. It is not known exactly when this change occurred.

One of two Type A "Tunnel Muckers" owned by F. Rolandi, San Francisco, California. The removal of the steam boiler, water tank, and coal bunk necessitated the stack of counterweights.

One of the first ERIE Type B shovels built. This machine was the one by which all others of its size were measured. First offered in 1915, 3,130 units were built before the Type B was replaced by the Type B-2 in 1925.

Type B owned by C. C. Beam of Yellow Springs, Arkansas. This machine was moving large pieces of limestone to the side, which would be shot again with dynamite. Those that did not fit in the bucket were lifted with chains.

Type B shovel equipped with wooden lagging on the traction wheels, which enabled the machine to travel across soft ground without the use of planking. The Ball Engine Company wisely decided to keep its product line narrow and produced only two models, the ¾-cubic yard Type B and the ½-cubic yard Type A.

An early Type B with caterpillar-type mounting. This machine was tearing up cobblestone pavement for R. E. Cash of Cincinnati, Ohio. Notice the fireman's "tools of the trade" that hung on the open door.

Standard Type B owned by Edgar Brothers Company, Metuchen, New Jersey. Working in a clay pit, the shovel loaded material into small side-dump cars.

Type B shovel owned by Arthur A. Johnson Corporation, Long Island, New York. The ERIE was excavating rock on a road job near Parsippany, New Jersey. This machine was equipped with a special long dipper handle for stripping and trenching.

"This is the hardest job I've ever had—and the best shovel!" claimed Edward Carson, the shovel operator. "I have worked on practically every make of shovel, but there isn't a one that can touch the ERIE!" This machine was owned by Sheesley & Janney, Johnstown, Pennsylvania.

The Thomasville Stone & Lime Company used four Type B machines in their operation. Three shovels worked at the rock face, while on top a Type B traction crane with a clam bucket stripped overburden.

A scene that was common at many construction sites—an ERIE Type B shovel loading Mack AC "Bulldog" dump trucks. When seeing a steam shovel for the first time, people often remarked about the "smoke" that came from the end of the boom. It was actually the steam exhaust from the crowd engine.

An early Type B highlift shovel owned by the Department of Public Works, Richmond, Virginia. The Ball Engine Company sold off its line of stationary engines in 1920. Two years later, to better associate it with its principal product, the company's name was changed to the Erie Steam Shovel Company.

Type B traction shovel with wooden lagging on traction wheels. The lagging consisted of oak blocks bolted into channel irons that were, in turn, bolted to the wheels. This increased tread width by 12 inches and, thus enabled the machine to traverse muddy or sandy terrain without the use of planking.

Five Type B shovels excavated for the Sears-Roebuck Building in Los Angeles, California. Under ideal working conditions, an ERIE could load two to three ¾-cubic yard dippers per minute. Since most early dump trucks could handle about two cubic yards, imagine the number of trucks it took to keep five shovels busy!

Late model ERIE Type B high lift shovel equipped with new type 19-foot 9-inch boom with heavier base and 16-foot dipper handle being tested at the factory. A note on the back of this photo states, "Two-speed propelling mechanism, operator's seat, running boards, trip rope idler, and two-way exhaust are now furnished as standard equipment on all Type B machines."

This Type B drew a big crowd. "Shovel watching" almost certainly started with the very first steam shovel, built in 1836 by William Otis. Note that the upper works were mounted forward of center on the crawler base, which balanced the machine and allowed the operator to bring the bucket in closer to the machine to start the cut.

ERIE Type B on a stripping operation for McCoy & Operman, Connellsville, Pennsylvania. The special long dipper handle enabled the operator to dispose of the material at a longer radius.

Standard Type B shovel owned by Boggs, Burnam & Company, Richmond, Kentucky. This machine was working at a 12-foot ledge of rock at the bottom of the quarry. Loading side-dump railroad cars, they averaged 600 cubic yards of rock per 10-hour day.

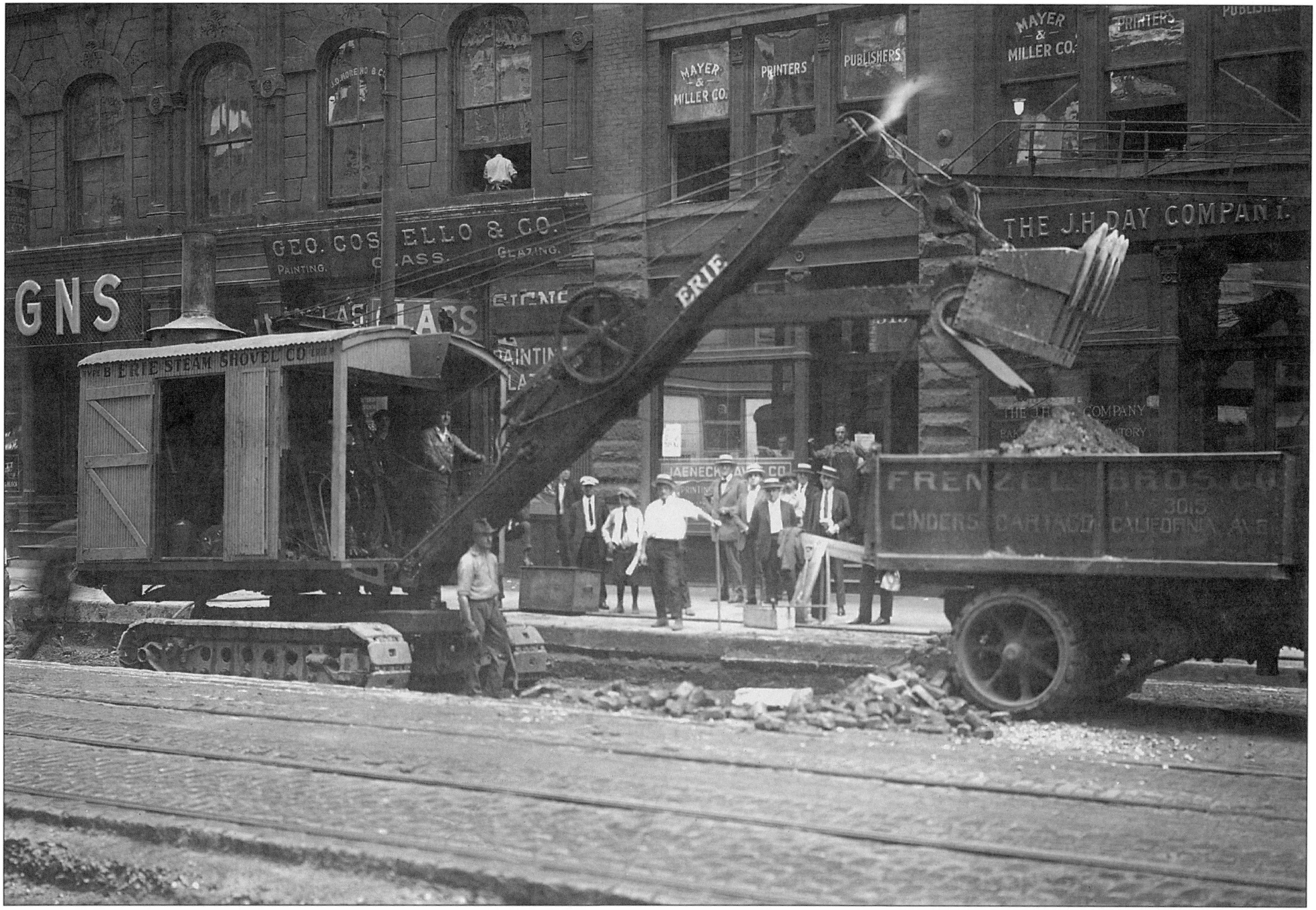

Type B ERIE caterpillar-type shovel owned by John O'Gara Company of Chicago, Illinois. This machine operated under extreme handicaps. Streetcar traffic was maintained, telegraph poles lined the curbs, and the material, brick pavement laid on a concrete base, had to be loaded in front of the machine.

Backfilling a trench with a Type B crane owned by Peter F. Connolly Company of Kent, Ohio. The small "trolley" on top of the boom was the traveling counterweight for the clam bucket tagline.

Type B crane on caterpillar-type mounting excavated gravel in Lamb, Michigan. The owner was M. P. Shanahan. Because of low single line pull, it was sometimes necessary to use a two-part closing line on the clam bucket.

Type B crane owned by William P. McDonald Construction Company, New York, New York. The machine was loading sand and stone into an Erie (different company) Aggremeter plant.

This was the first caterpillar-type machine ERIE built. Owned by the Moore-Young Construction Company of Waterloo, Iowa, the machine was purchased in 1921 and used for loading sand into a hopper.

Erie Type B crane on a road job in Detroit, Michigan. Owned by the T. E. Currie Company, it unloaded aggregate gravel directly from 50-ton hopper cars into the loading bin. A small Plymouth industrial locomotive pulled the loaded cars directly to the concrete paving spread.

Equipped with a one-cubic yard orange peel bucket, a Type B crane built a breakwater at White's Point for C. R. Butterfield of San Pedro, California. The rock had been drilled and blasted at low tide, and the ERIE worked until the tide would get too high.

T. J. M. Contracting Company of Brunswick, New Jersey, had their Type B crane unloading cars of sand and stone for road construction into steel measuring bins and storage piles.

Type B crane equipped with a combination 36-foot and 40-foot crane boom. Set for clamshell work, the tip sheave could easily be moved out to the end for the longer reach desirable for dragline work or for use with a clambucket smaller than ¾ cubic yard.

Early Type B traction crane owned by Sweeney Brothers, Scranton, Pennsylvania. This machine loaded material directly from the bank into rail cars. In the background, an ERIE Type B traction shovel performed the same task.

Basement excavation for the Biltmore Hotel in East Providence, Rhode Island. The Joseph McCormick Company dug the hole as deep as possible with ERIE shovels. Once they could no longer reach the trucks above, one machine was converted to a crane and finished the deep digging with a clam bucket.

Type B crane owned by the Perry-Buxton-Doane Company of Portland, Maine, dealers in iron and steel scrap. The crane was equipped with a 30-inch diameter Ohio magnet, which weighed 3,400 lbs. Power for magnet operation was supplied via lead cord to a motor generator set in the center of the property.

Type B crane owned by T. E. Currie Company of Detroit Michigan. This clamshell excavated a pier for a new bridge. In the background was a floating pile driving outfit, consisting of a derrick and drop hammer and a steam donkey engine for power.

Appearing on this page and in two photos opposite is a Type B dragline that was owned by Odell Brothers, Fort Collins, Colorado. A three-part hoist line and two-part dragline demonstrated the low line pull exerted by most steam cranes of comparable size during this era. Digging at an average depth of three feet, this machine typically managed 300 lineal feet of ditch per 10-hour day.

In this area, the ditch ran deeper than three feet. Note the supply wagons in the background. A water wagon and a wagon with sideboards for hauling coal kept the ERIE steamed up.

The ditch averaged 12 feet across at the top; three feet at the bottom. It is interesting to note that the spoil was being cast to both sides of the excavation.

Factory view of dragline fairlead arrangement on a Type B crane. Four rotating sheaves insured that the drag cable would not flatten and wear out prematurely.

Type B dragline, owned by the State of South Dakota, Department of Game and Fish, dug a drainage ditch near Dolton, South Dakota.

The State of South Dakota machine at a closer range. Steering the ERIE traction machines was a simple affair. A link bar was attached between the upper works and the steering front axle. By slightly rotating the house, the operator accurately steered the crane.

Type B shovels could be specially equipped for deep trench work. Here, the traction wheels were removed. The truck frame was mounted to a structural steel, wide gauge track supplied by ERIE and was suspended over the excavation.

A similar setup to that on the previous page used a crawler-mounted Type B shovel. An extension was added to the 22-foot long dipper handle. The shovel, owned by E. R. Moorehouse Company of Camden, New Jersey, backfilled its own work as it went, loading the "swell" into bottom-dump wagons.

The Moorehouse shovel also set the 7-foot 6-inch-diameter pipe, which weighed seven tons. The average day's work competed was 30 lineal feet of sewer, with 50 lineal feet completed on the most productive day. The owner claimed to have cleared between $600 and $700 per day, enough to have paid for the machine after 20 days of operation.

The inherent disadvantages of using a shovel above a trench eventually encouraged shovel manufacturers to develop a new excavator, the backhoe. ERIE's first hoe, built in 1921, was simply devised by reversing the trenching shovel bucket at the end of the dipper. In this photo, the machine, owned by Columbia Construction Company of Indianapolis, Indiana, excavated at a depth of 14 feet.

The Columbia Construction "steam hoe" had excavated 50 lineal feet in 4 ½-hours, the day that this photograph was taken. Note the temporary shoring used to hold back the embankments.

In a more common ditching arrangement, a Type B trench shovel sat on timber mats that straddled the trench. The machine was “moved up” by swinging the mats around to the front of the machine before another cut was made. Note the limited dumping height, when the machine was equipped with the long dipper handle. The boom suspension “hog rods” indicate that this machine was not equipped with an independent boom hoist.

The F. A. Frobose Company of Lima, Ohio used a Type B trench hoe to construct a concrete pipeline for the City of Lima. An ERIE crane followed along to set the 78-inch diameter, 11,000 lb. reinforced concrete pipe.

An illustration of the underside of a Type B traction mounting.

A view of the Type B caterpillar-type mounting, with impoved steering arrangement. The operator would drop a tiller bar from the upper works into the pocket in the front center of the undercarriage and, by slightly swinging the house right or left, could engage and disengage each track independently.

This photo shows the tiller bar in position.

A view of the special ERIE manganese front dipper, recommended for certain classes of rock excavation. The steel plate on the dipper sides was ⅝ inches thick. The front measured one inch thick.

Type B shovel buckets were rated at ¾ cubic yard water measure capacity. This special "quarry dipper" had a manganese front that measured 1 ¼ inch thick. The bucket was wider at the back than at the front and, thus, rocks were prevented from becoming wedged in the bucket.

A view of the underside of the Type B machinery deck. The massive casting insured the rigid alignment of machinery and, thus, resulted in a long life for components.

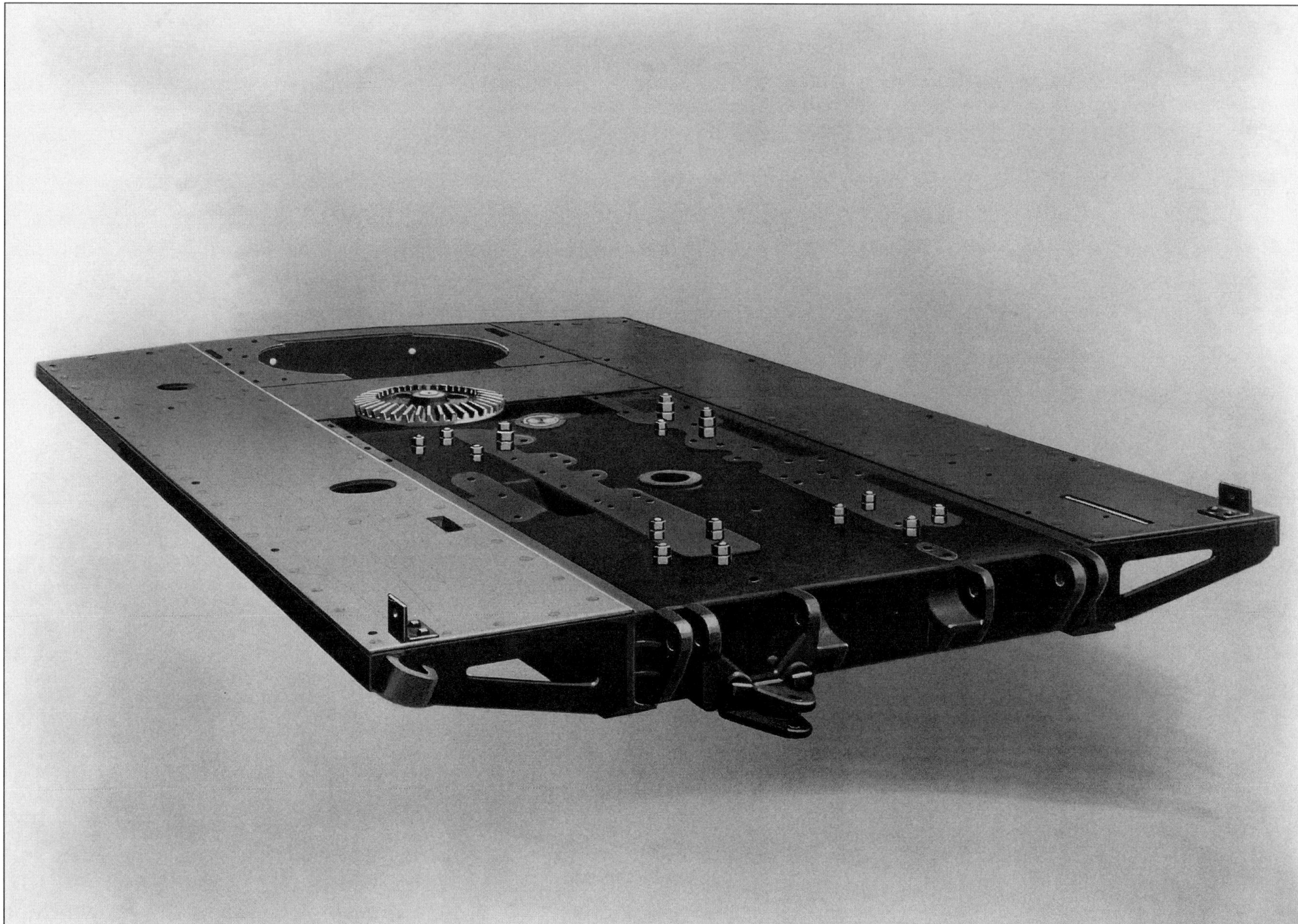

A view of the top of the Type B machinery deck. All mounting areas were machined and planed to close tolerances on jigs, which insured the perfect alignment of original and replacement components.

A front view of the layout of the Type B deck machinery. The design of this ERIE machine was so perfect that only minor improvements were made over the course of production.

Side view of the operator's station and deck machinery layout of the Type B. The three lever arrangement was simple and efficient. ERIE machines were nearly effortless in operation, as the clutches were steam set. The first movement of the control lever set the clutch; further motion opened the throttle and activated the machinery; complete movement fully opened the throttle.

Back side view of the vertical, double-reversing hoisting engines used on the Type B.

Boilers used on the Type B shovels were of the vertical, multitubular-type. Tested to 222 lbs., the boiler was designed for a working pressure of 148 lbs. However, a working pressure of 125 to 135 lbs. was more common.

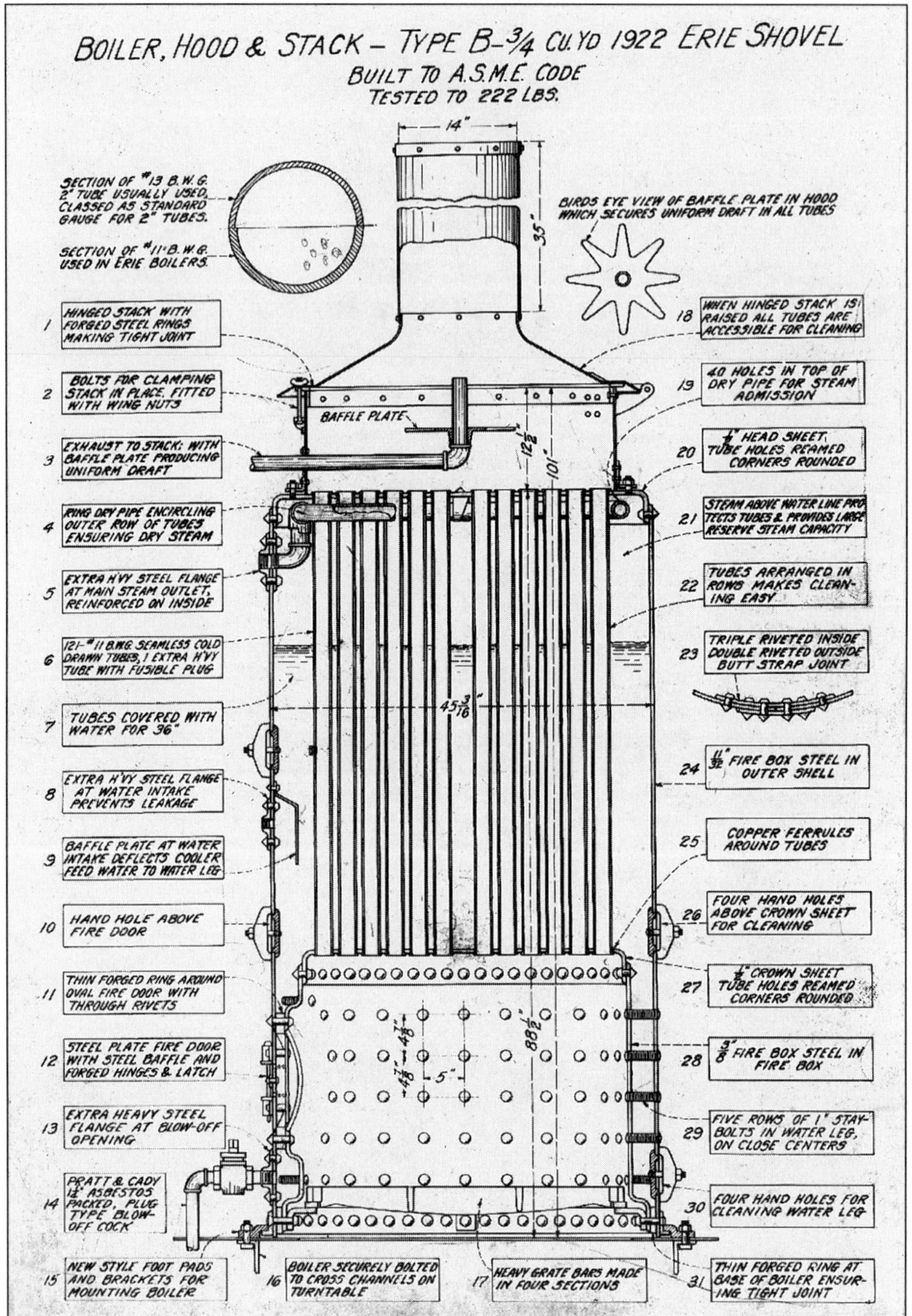

A drawing of the Type B boiler, hood, and stack.

A crowd engine with the dust cover removed. The crowd and swing reversing engines were nearly identical. The significant difference between them was in the valve control. When the crowd engine was "on center", the dipper handle would hold its position. When the swing engine was "on center", the control valve allowed the upper works to drift or free-swing.

A view of the front of a Type B shovel. The boom was constructed of select white oak and clad with steel channel reinforcement. The oak provided the boom with the flexibility needed for the strains of shovel work.

A rear view of the house of a Type B shovel. Access doors were for boiler maintenance and coal storage.

A side view of a Type B house. The boiler stack was hinged, to provide additional height clearance during transport. Also in view is the worm-driven, self-locking boom hoist.

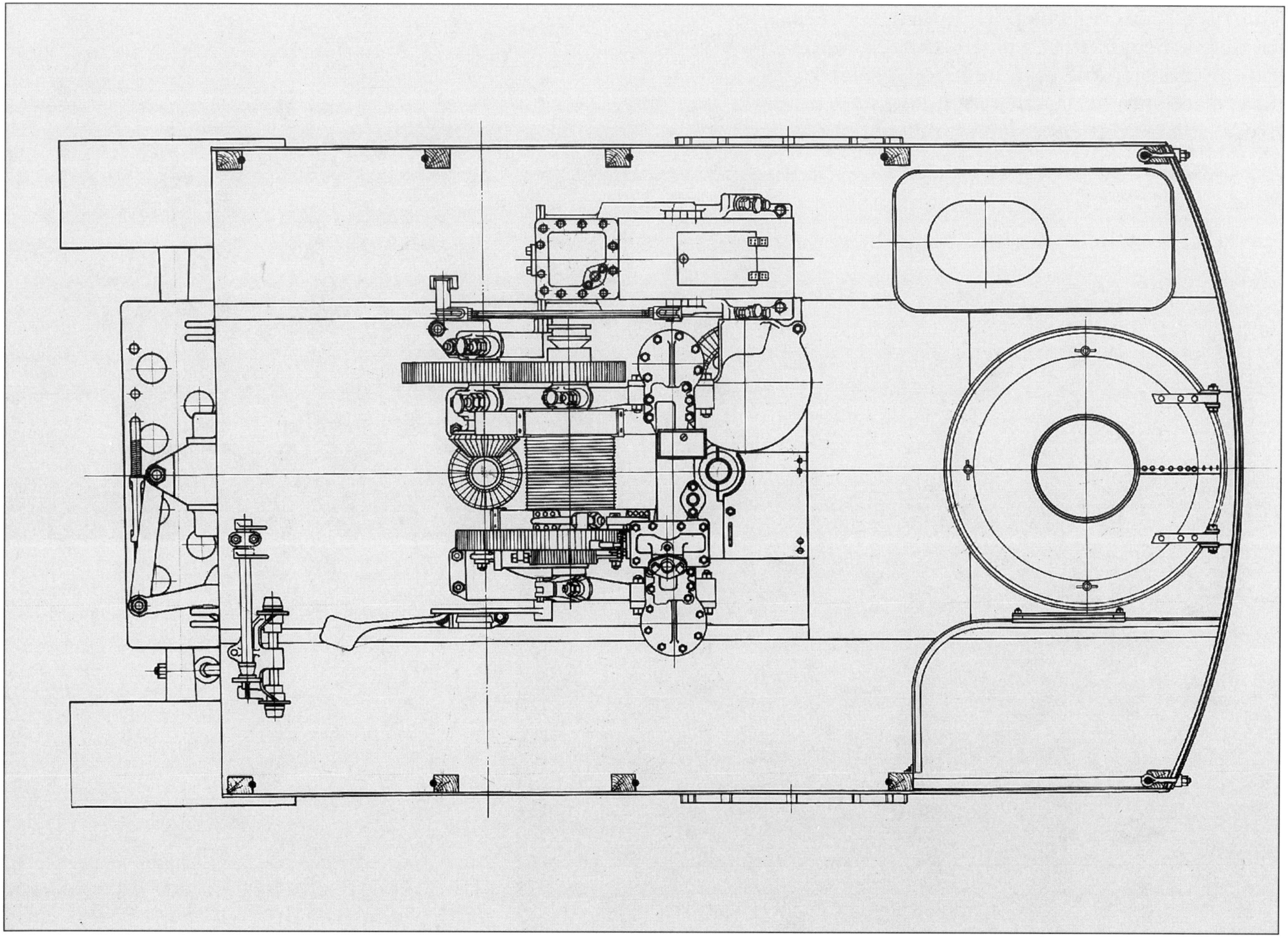

A deck machinery plan for a Type B shovel. Coal bunk was located to the left of the boiler; water tank at the right.

The Type B shovel was particularly successful as a railroad "ditcher". By simply changing the truck frame on a high-lift shovel and adding a boom hoist, the machine was ready to operate from the top of a standard flat car.

The Type B ditcher could traverse the entire length of the flat car, and was thus able to load material into cars positioned at either end.

Ditchers were used to clear slides and for excavating and cleaning ditches. A.H.&D. built the first ditcher in 1905. The Marion Steam Shovel also offered its Model 21 and 31 as popular ditchers. The ERIE Type B ditcher in this photo was loading into 20-cubic yard air-dump cars.

An underside view of the Type B railway ditcher "racer". Note that the wheels were set for three different rail gauges.

Top view of a Type B railway ditcher truck. The two outriggers were used to maintain position and to improve stability on the flat car.

The ERIE Type B-2 Dreadnaught. By 1924, ERIE management recognized the need for a machine larger than the ¾-cubic yard Type B. In 1926, the Type B was dropped from production and the B-2, a one-cubic yard machine, replaced it.

The new Type B-2 Dreadnaught in action on cellar excavation. The Dreadnaught's 25 ton weight was a substantial increase over the 20 tons of the Type B.

Three Type B-2 cranes on J. J. Dunnegan Construction Company's job in South Bend, Indiana. Two machines drove 33-foot sheet piling with McKiernan-Terry steam hammers that weighed 5,500 lb. These ERIE's also had special 54-inch by 100-inch boilers that supplied steam to the hammers.

The third J. J. Dunnegan Construction Company ERIE machine on the job setting the sheet piling. The Type B-2 was equipped with a 50-foot box-type boom and an additional 40-foot long 12-inch by 12-inch timber that increased the boom length to 75 feet.

A new ERIE Type B-2 Dreadnaught shovel. This machine may have been built after the ERIE merger with Bucyrus in 1927, as it featured improved chain drive-type crawlers.

A new Type B-2 steam hoe as it was "run in" behind the plant in Erie, Pennsylvania. By this time, ERIE had adopted the more common and successful backhoe arrangement popularized by Northwest Engineering Company, Green Bay, Wisconsin, one of ERIE's chief rivals.

A beautiful view of a new ERIE Type Gas+Air or Type GA-2 shovel owned by the Navarro Company, Pittsburgh, Pennsylvania. ERIE first offered the Gas+Air shovel very late in 1925. The company was slow in moving away from steam power and to use of the internal combustion engine, as had so many of its competitors.

The unique Gas+Air used compressed air-powered engines very similar to the steam engines used on ERIE steamers. The crowding and swinging functions were performed by air pressure. Hoisting, booming, and traveling were powered mechanically.

A fine job of street grading was done by the Type Gas+Air owned by Henry Bickle Company, Louisville, Kentucky. Power was supplied by a Waukesha Model WS four-cylinder gasoline engine.

On this three-foot road grading cut, with stiff shale bottom, the Gas+Air loaded at the rate of 22 four-cubic yard trucks per hour. With the ERIE's one-cubic yard bucket, average loading time was one minute, fifteen seconds.

This Type GA-2 shovel made good time as it dug up an old road for a new concrete highway, even though, in order to load trucks, it had to make a 180-degree swing.

Gas+Air, owned by Geo. M. Brewster & Son, Bogota, New Jersey, as it dug in a high bank on a road job near Park Ridge, New Jersey. The compressed air was heated and, thus, delivered more air per cubic foot minute.

Gas+Air as it broke, without blasting, mid-winter frost more than one foot thick. The machines compressed air crowding motion did not rob power from its mechanical hoisting motion. This was a big advantage that Gas+Air had over conventional gas engine-powered shovels.

Gas+Air, owned by Aris-Knapp Company, Oakland, California, as it dug a deep cellar, tearing out old foundation. The primary complaints about the Gas+Air were the amount of maintenance it required and the difficulty encountered in making the "high-tech" adjustments necessary to this hybrid machine.

This Gas+Air backhoe used the same hoe front as did the Type B-2 Dreadnaught. ERIE cranes required the installation of a second hoist drum, seen positioned at the base of the boom.

The Gas+Air used some components from the ERIE steamers. The box-type boom and crawler mounting were two examples. This one cubic yard-machine was also equipped with a boom-mounted dragline fairlead.

A view of a Type GA-2 shovel without the house installed. The machinery arrangement was much more cluttered than that of the ERIE steamers. The compressed air passed through the exhaust pipe and, thus, was expanded by the heat.

An overhead view of a Type GA-2 shovel. The Waukesha engine was mounted directly above the air compressor. The "Little David" air compressor is visible just ahead of the fuel tank.

A rear view of the power plant and air compressor used in the Gas+Air machine. Also visible is the fuel tank, to the left, and air receiver, to the right.

An aerial view of a completely assembled turntable from a Gas+Air shovel. Note the easy accessibility to the motor and swing engines, and the McCord Lubricator, used to lubricate the moving parts of the compressed air engines. Hoisting clutches were air set, which resulted in the need for little effort to operate the GA-2.

An interesting view of the rear of the deck of a GA-2 machine. The 80 horsepower, 5 ¾-inch by 8-inch Waukesha WS engine was hinged to the air compressor below. This enabled adjustment or replacement of the connecting rod bearings without removal of the engine.

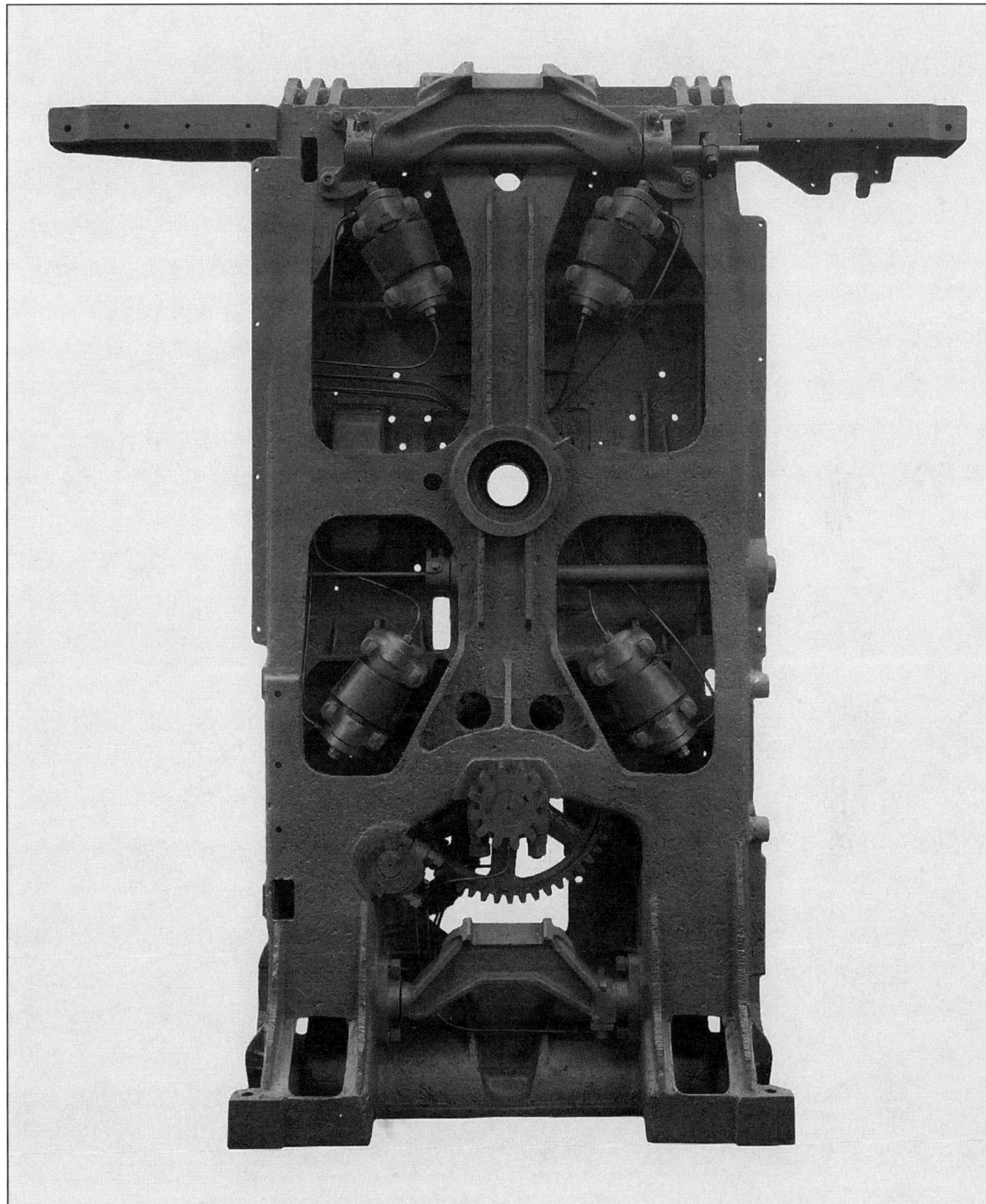

A view of the underside of a Type GA-2 turntable. Erie claimed that their turntable was four times stronger than those of competitors.

A GA-2 shovel shipped completely assembled on one flat car.

An ERIE GA-2 shovel with its cab fully opened, a feature appreciated by its operator during warm weather. The Gas+Air four-cylinder Waukesha engine was crank-started but, after building air, it could be shut down and restarted with its "Little David" air starter. This machine sported the greatly-improved chain-driven crawlers.

Erie Shovel Shipments 1915 - 1927

Yard	Total	Type A	Type B	Type B2	Type GA2	Type DA2
1915	50	-	50	-	-	-
1916	137	-	137	-	-	-
1917	213	-	213	-	-	-
1918	189	7	182	-	-	-
1919	321	66	255	-	-	-
1920	446	54	392	-	-	-
1921	297	72	225	-	-	-
1922	529	78	451	-	-	-
1923	576	78	498	-	-	-
1924	440	72	368	-	-	-
1925	425	54	359	3	9	-
1926	487	34	-	337	116	-
1027	329	14	-	212	102	1

Photo Credits

PAGE 8 Keith Haddock **PAGE 21** Keith Haddock **PAGE 30** Keith Haddock **PAGE 31** Keith Haddock **PAGE 32** Keith Haddock **PAGE 34** Keith Haddock **PAGE 49** Keith Haddock **PAGE 50** Keith Haddock **PAGE 96 LOWER LEFT** Keith Haddock **PAGE 98** Ron Ketron **PAGE 99** Ron Ketron **PAGE 102** Ron Ketron **PAGE 103** Ron Ketron **PAGE 104** Ron Ketron **PAGE 105** Ron Ketron **PAGE 106** Keith Haddock **PAGE 107** Keith Haddock **PAGE 108** Keith Haddock **PAGE 109** Keith Haddock **PAGE 110** Keith Haddock **PAGE 111** Keith Haddock **PAGE 112** Keith Haddock **PAGE 113** Keith Haddock **PAGE 114** Keith Haddock **PAGE 115** Keith Haddock **PAGE 116** Ron Ketron **PAGE 117** Ron Ketron **PAGE 118** Ron Ketron **PAGE 119** Keith Haddock **PAGE 120** Keith Haddock **PAGE 123** Ron Ketron

The Iconografix Photo Archive Series includes:

AMERICAN CULTURE

AMERICAN SERVICE STATIONS 1935-1943 ISBN 1-882256-27-1
COCA-COLA: A HISTORY IN PHOTOGRAPHS 1930-1969 ISBN 1-882256-46-8
COCA-COLA: ITS VEHICLES IN PHOTOGRAPHS 1930-1969 ISBN 1-882256-47-6
PHILLIPS 66 1945-1954 ISBN 1-882256-42-5

AUTOMOTIVE

IMPERIAL 1955-1963 ISBN 1-882256-22-0
IMPERIAL 1964-1968 ISBN 1-882256-23-9
LE MANS 1950: THE BRIGGS CUNNINGHAM CAMPAIGN ISBN 1-882256-21-2
PACKARD MOTOR CARS 1935-1942 ISBN 1-882256-44-1
PACKARD MOTOR CARS 1946-1958 ISBN 1-882256-45-X
SEBRING 12-HOUR RACE 1970 ISBN 1-882256-20-4
STUDEBAKER 1933-1942 ISBN 1-882256-24-7
STUDEBAKER 1946-1958 ISBN 1-882256-25-5
LINCOLN MOTOR CARS 1920-1942 ISBN 1-882256-57-3
LINCOLN MOTOR CARS 1946-1960 ISBN 1-882256-58-1
MG 1945-1964 ISBN 1-882256-52-2
MG 1965-1980 ISBN 1-882256-53-0
GT40 ISBN 1-882256-64-6
FERRARI PININFARINA 1952-1996 ISBN 1-882256-65-4
VANDERBILT CUP 1936 & 1937 ISBN 1-882256-66-2

TRACTORS AND CONSTRUCTION EQUIPMENT

CASE TRACTORS 1912-1959 ISBN 1-882256-32-8
CATERPILLAR MILITARY TRACTORS VOLUME 1 ISBN 1-882256-16-6
CATERPILLAR MILITARY TRACTORS VOLUME 2 ISBN 1-882256-17-4
CATERPILLAR SIXTY ISBN 1-882256-05-0
CATERPILLAR THIRTY ISBN 1-882256-04-2
CLETRAC AND OLIVER CRAWLERS ISBN 1-882256-43-3
ERIE SHOVEL ISBN 1-882256-69-7
FARMALL F- SERIES ISBN 1-882256-02-6
FARMALL MODEL H ISBN 1-882256-03-4
FARMALL MODEL M ISBN 1-882256-15-8
FARMALL REGULAR ISBN 1-882256-14-X
FARMALL SUPER SERIES ISBN 1-882256-49-2
FORDSON 1917-1928 ISBN 1-882256-33-6
HART-PARR ISBN 1-882256-08-5
HOLT TRACTORS ISBN 1-882256-10-7
INTERNATIONAL TRACTRACTOR ISBN 1-882256-48-4
JOHN DEERE MODEL A ISBN 1-882256-12-3
JOHN DEERE MODEL B ISBN 1-882256-01-8
JOHN DEERE MODEL D ISBN 1-882256-00-X
JOHN DEERE 30 SERIES ISBN 1-882256-13-1
MINNEAPOLIS-MOLINE U-SERIES ISBN 1-882256-07-7
OLIVER TRACTORS ISBN 1-882256-09-3
RUSSELL GRADERS ISBN 1-882256-11-5
TWIN CITY TRACTOR ISBN 1-882256-06-9

RAILWAYS

CHICAGO, ST. PAUL, MINNEAPOLIS & OMAHA RAILROAD 1880 - 1940 ISBN 1-882256-67-0
GREAT NORTHERN RAILWAY 1945-1970 ISBN 1-882256-56-5
MILWAUKEE ROAD 1850-1960 ISBN 1-882256-61-1
SOO LINE 1975-1992 ISBN 1-882256-68-9

TRUCKS

BEVERAGE TRUCKS 1910-1975 ISBN 1-882256-60-3
*BROCKWAY TRUCKS 1948-1961** ISBN 1-882256-55-7
DODGE TRUCKS 1929-1947 ISBN 1-882256-36-0
DODGE TRUCKS 1948-1960 ISBN 1-882256-37-9
LOGGING TRUCKS 1915-1970 ISBN 1-882256-59-X
*MACK® MODEL AB** ISBN 1-882256-18-2
*MACK AP SUPER DUTY TRUCKS 1926-1938** ISBN 1-882256-54-9
*MACK MODEL B 1953-1966 VOLUME 1** ISBN 1-882256-19-0
*MACK MODEL B 1953-1966 VOLUME 2** ISBN 1-882256-34-4
*MACK EB-EC-ED-EE-EF-EG-DE 1936-1951** ISBN 1-882256-29-8
*MACK EH-EJ-EM-EQ-ER-ES 1936-1950** ISBN 1-882256-39-5
*MACK FC-FCSW-NW 1936-1947** ISBN 1-882256-28-X
*MACK FG-FH-FJ-FK-FN-FP-FT-FW 1937-1950** ISBN 1-882256-35-2
*MACK LF-LH-LJ-LM-LT 1940-1956 ** ISBN 1-882256-38-7
*MACK MODEL B FIRE TRUCKS 1954-1966** ISBN 1-882256-62-X
*MACK MODEL CF FIRE TRUCKS 1967-1981** ISBN 1-882256-63-8
STUDEBAKER TRUCKS 1927-1940 ISBN 1-882256-40-9
STUDEBAKER TRUCKS 1941-1964 ISBN 1-882256-41-7

The Iconografix Photo Archive Series is available from direct mail specialty book dealers and bookstores worldwide, or can be ordered from the publisher. For additional book trade and distribution information or to add your name to our mailing list contact:

Iconografix
PO Box 609/BK
Osceola, Wisconsin 54020 USA

Telephone: (715) 294-2792
(800) 289-3504 (USA)
Fax: (715) 294-3414

Book trade distribution by Voyageur Press, Inc., PO Box 338, Stillwater, Minnesota 55082 USA (800) 888-9653
European distribution by Midland Publishing Limited, 24 The Hollow, Earl Shilton, Leicester LE9 7N1 England